Housing 229

去火星生活？

Life on Mars?

Gunter Pauli

[比] 冈特 · 鲍利 著

[哥伦] 凯瑟琳娜 · 巴赫 绘

李原原 译

上海远东出版社

特别感谢以下热心人士对童书工作的支持：

匡志强　方　芳　宋小华　解　东　厉　云　李　婧
刘　丹　熊彩虹　罗淑怡　旷　婉　杨　荣　刘学振
何圣霖　王必斗　潘林平　熊志强　廖清州　谭燕宁
王　征　白　纯　张林霞　寿颖慧　罗　佳　傅　俊
胡海朋　白永喆　韦小宏　李　杰　欧　亮

目录

去火星生活? 4

你知道吗? 22

想一想 26

自己动手! 27

学科知识 28

情感智慧 29

艺术 29

思维拓展 30

动手能力 30

故事灵感来自 31

Contents

Life on Mars? 4

Did You Know? 22

Think about It 26

Do It Yourself! 27

Academic Knowledge 28

Emotional Intelligence 29

The Arts 29

Systems: Making the Connections 30

Capacity to Implement 30

This Fable Is Inspired by 31

一个小女孩正仰望天空。爷爷注意到她眼中的光芒，问她：

“你认为火星上有生命吗？”

“哦，爷爷，很高兴您问我这个问题。那您说说，您认为火星上有生命吗？”

“我认为那儿有。我们不可能是这未被探索的广袤宇宙中唯一的生命。”

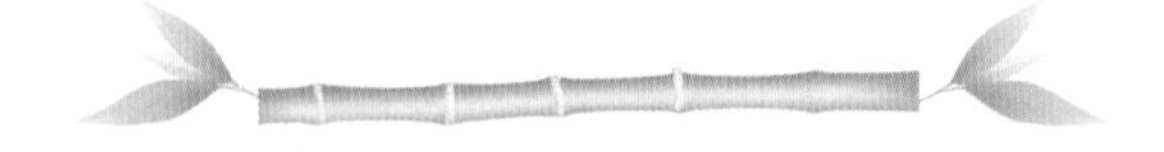

A little girl is watching the sky. Her grandfather notices the sparkle in her eyes and asks her:

“Do you think there is life on Mars?”

“Oh Granddad, I am so happy you’re asking me this. Tell me, do you think there is life on Mars?”

“I am convinced that there is life out there. It cannot possibly be that we are the only ones alive in this vast unexplored Universe.”

一个小女孩正仰望天空。

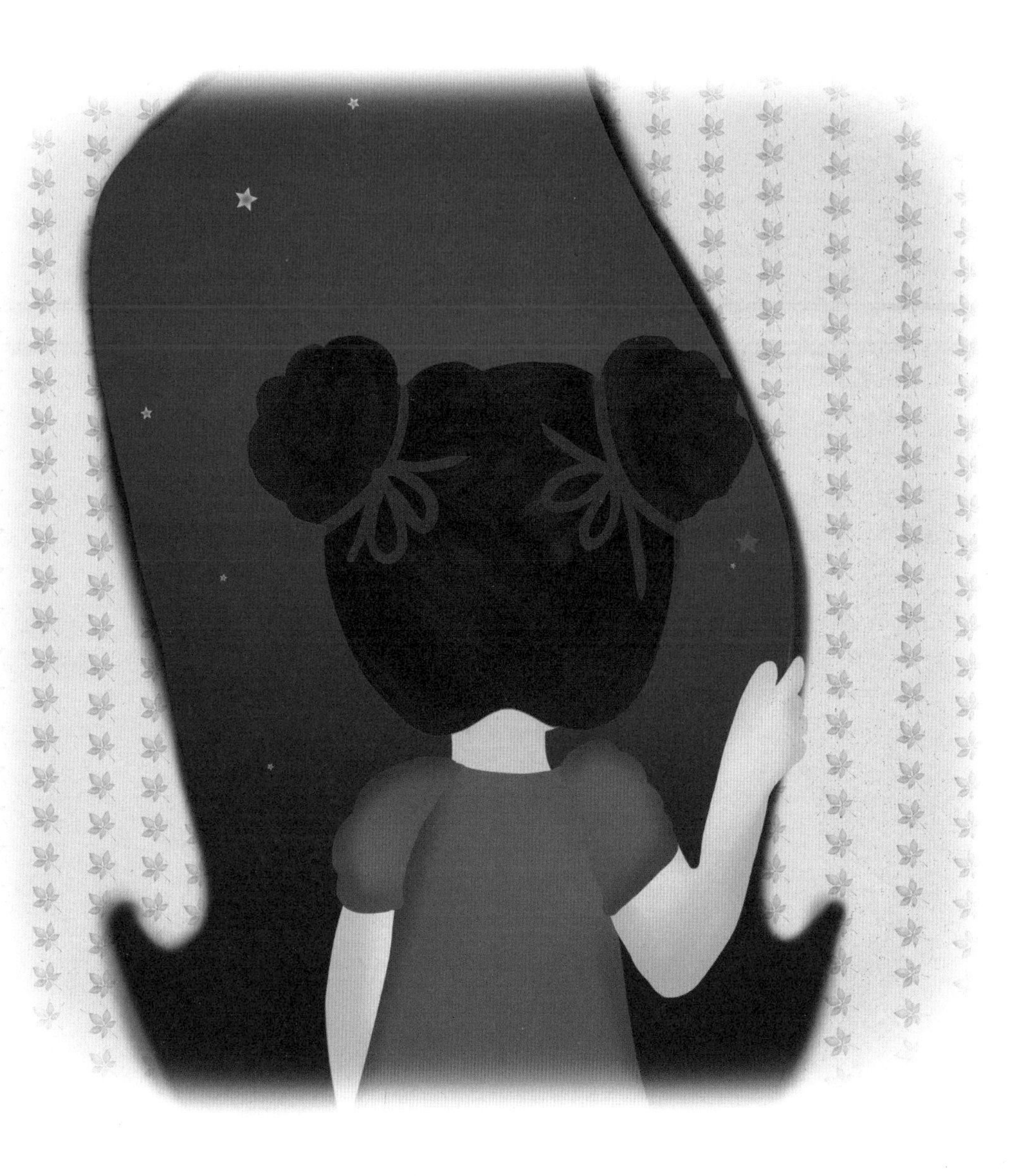

A little girl is watching the sky.

我非常非常想去那儿……

I do dream of going out there ...

“我非常非常想去那儿探索新世界。现在看来我们的地球已经没有什么新东西了。”

“但是太空太大、太遥远了……如果你去那儿，我可能就再也见不到你了！”

“但这不正是我们的祖先所做的吗？”

“嗯，当年你的曾曾曾曾祖父母为了活命逃离法国时，他们乘船去了南非……”

“I do dream of going out there and exploring new worlds. It seems that we are running out of new ideas here on Earth.”

“But Space is so vast … if you go there, I may never see you again!”

“But is that not exactly what our ancestors did?”

“Well, when your great-great-great-great-grandparents fled for their lives from France they got on a ship and sailed to South Africa …”

“他们对那个地方知之甚少，也知道自己永远回不了家了。所以我只是在延续家族传统，不是吗？”

“是的,亲爱的。但我会非常想念你。我知道我可能再也见不到你了，看着你离去，我的心都要碎了。”

“我也会想念你的，爷爷。但是想象一下！我将去创造生命——在一个我们现在认为不存在生命的地方。”

“To a place they knew very little about, knowing they would never return home. So I would just be continuing a family tradition, wouldn't I?”

“True, my dear. But I would miss you terribly, and it would break my heart to see you go off, knowing I'd probably never see you again.”

“And I would miss you too, Granddad. But just imagine! I would go to create life – in a place we currently believe there not to be any.”

我会非常想念你。

I would miss you terribly.

你真聪明！

You are such a smart girl!

“这听起来像是‘不可能完成的任务’。”爷爷说。

“然而，不可能和不太可能之间有很大的区别。我不认为这是不可能的，所以……我需要的是正确的科学知识、创新和丰富的想象力。”

“你真聪明！无论你想象中的是什么，只要你掌握了其中的科学知识，就很有可能将幻想变成现实。”

“That sounds like a ‘mission impossible’,” Granddad comments.

“There is a big difference between what is impossible and what is improbable, though. I don’t consider it impossible, so … what I will need is sound science, innovation and a lot of imagination.”

“You are such a smart girl! Whatever pops up in your imagination may very well turn from fantasy into reality – provided you grasp the science behind it.”

“但不是所有的科学知识，爷爷！如果你的科学知识告诉我这是不可能的，那么我就会去发现、学习和分享那些认为这是可能的科学知识。”

“真棒，通过探索，我们已经完成了看似不可能的事情。比如，我们发现了比你小脚趾的十亿分之一还小的东西，发现了比你一根头发的万亿分之一还细的东西。”

“您太智慧了！我们确实需要更多地了解像分子、原子这样的小东西，还需要了解像行星、恒星和星系那样的大东西。我们应该时刻准备着去发现那些不可想象的东西。”

“But not just any science, Granddad! If your science tells me it is impossible, then I want to discover, learn and share the science that says it is indeed possible.”

“Well, we have already achieved the seemingly impossible by discovering things that are smaller than a billionth of your little toe, or a trillionth of the thickness of one of your hairs.”

“You are so wise! We do indeed need to know more about the very small things, like molecules and atoms, as well as the big things like planets, stars and galaxies. We should always be ready to discover the unimagined.”

……像行星、恒星、星系那样的。

... like planets, stars and galaxies.

……那里有一些水。

... there is some water.

“是的，还有一个人的态度。”爷爷补充道。“看来你有决心解决前人尚未解决的问题。”

“我当然喜欢寻找解决办法！这不就是我们学习的方式吗？我现在就有一个非常想解决的问题，那就是地球上的贫困问题。”

“嗯嗯，那你在火星上找不到贫困问题，或者其他任何东西。幸运的是，那里有一些水。”

“Then there is also, of course, one's attitude,” Granddad adds. “It seems you are determined to solve problems that have never been solved before.”

“I do love finding solutions! Isn't that how we learn? One problem I would really like to solve is that of people living in poverty on Earth.”

“Well, on Mars you won't find any poverty – or much of anything else, for that matter. Fortunately, there is some water.”

“并且在火星上，金钱将毫无价值。想象一下，当我们专注于拥有干净的水和种植健康的食物时，我们就会满足于我们所拥有的。”

“可是一开始，你将一无所有。你必须去创造你的生活所需，因为你不可能像现在一样从商店里买东西……”

“别担心，爷爷！如果你来看我时忘记带牙刷，我们会给你打印一把。”

“And on Mars, money will have no value. Just imagine: focusing on having clean water and growing healthy food, we'll be content with what we have.”

“To start with, you will have nothing. You will have to create everything you need, as there could be no quick visits to the shop…”

“Don't you worry, Granddad! Should you come visit and forget your toothbrush, we will just print you one.”

如果你来看我……

Should you come visit ...

……那我们总可以开采小行星吧。

... we can always mine the asteroids.

“如果我需要一根骨头或一台新电脑怎么办？你能打印吗？你从哪儿获取原材料？你不可能把地球上的一切都带走。”

“是的，那我们总可以开采小行星吧。”女孩建议道。

“And what if I need a bone, or a new computer? Will you be able to just print one? Where would you get your raw materials? It would be impossible for you to take everything with you from Earth.”

“Well, we can always mine the asteroids,” the girl suggests.

“开采小行星？这是多么强大的创造性思维和毅力啊！我真希望火星人欢迎这种想法，而不是为此生气……”

“请记住，亲爱的爷爷，所有我们为那个你认为什么都没有的火星所设想的解决方案，也同样适用于这个被我们挥霍浪费的地球。”

……这仅仅是开始！……

“Mining asteroids? Now, how is that for creative thinking, and perseverance! I do hope the Martians will welcome such thinking and not be annoyed by it …”

“Remember, my dear Granddad, that any solution we imagine for Mars where you think there is nothing, will help find solutions on Earth where we squander everything.”

... AND IT HAS ONLY JUST BEGUN!...

……这仅仅是开始！……

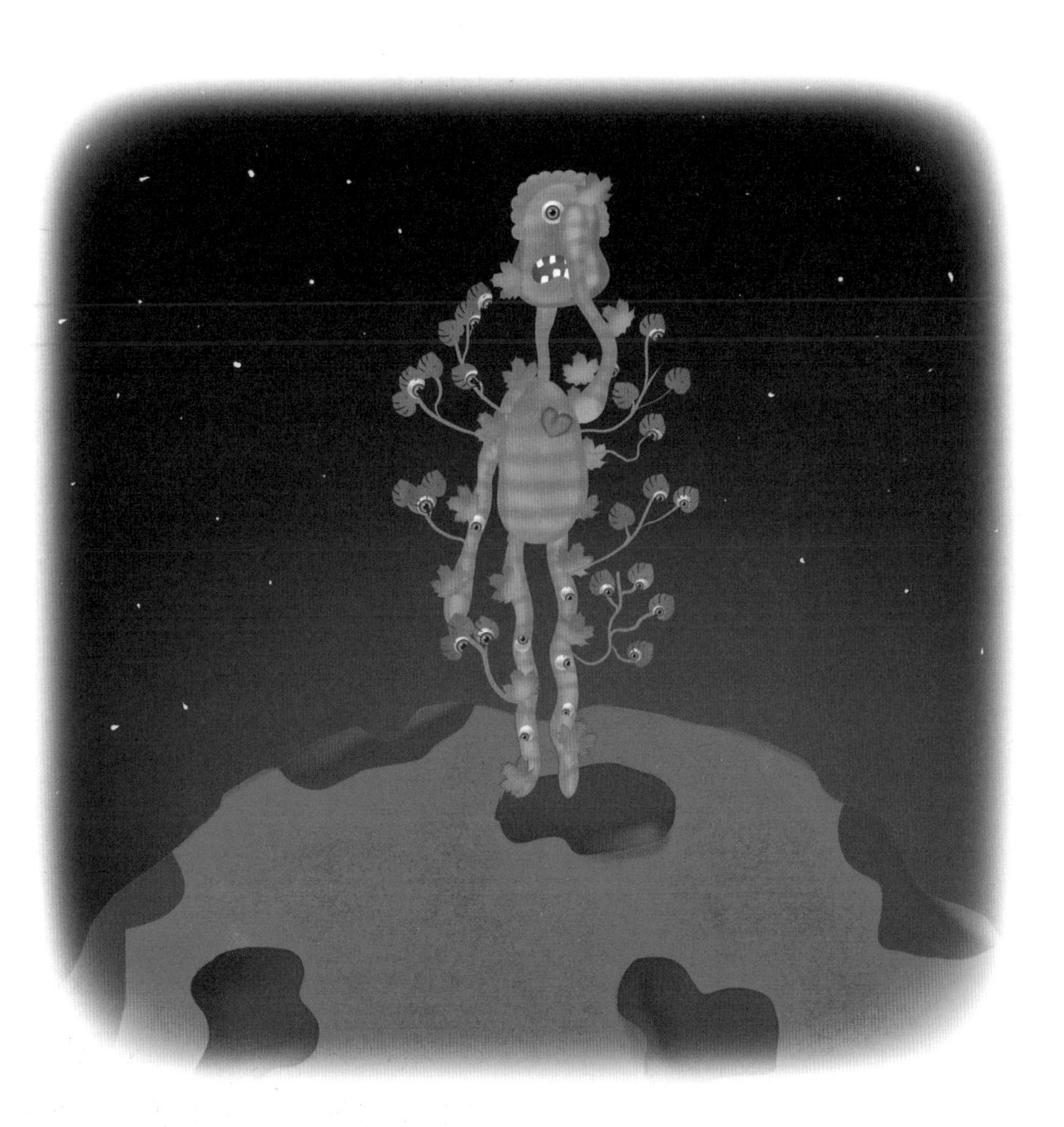

... AND IT HAS ONLY JUST BEGUN! ...

Did You Know?

你知道吗？

Scientists consider Mars a good destination for discovery and exploration. Its process of formation and evolution is comparable to that of Earth, and in the past Mars had conditions suitable for life as we know it.

科学家们认为火星是一个值得探索的好地方。它的形成和进化过程与地球相当，而且正如我们所知，火星曾经有适合生命生存的条件。

Mars has been studied by way of robotic exploring for more than 40 years. First studies began in low-Earth orbiting, aboard the International Space Station, where astronauts have researched and developed food, health and communication systems necessary for missions into deep Space.

40 多年来，人们一直通过机器人探索来研究火星。最初的研究开始于近地轨道，在国际空间站上，宇航员们已经研究和开发了进入外太空执行任务所必需的食品、健康和通信系统。

一批像“好奇号”探测器这样的航天器和火星车收集了火星及其周围的数据。火星表面含有氧化铁（铁锈），因此火星表面有了现在这样的颜色。火星大气中 95% 是二氧化碳，平均温度是零下 60℃。

A fleet of robotic spacecraft and rovers, like the Curiosity Rover, collects data on and around Mars. Its surface contains iron oxide (rust), giving it its colour. The atmosphere is 95% CO_2 and the average temperature is -60 °C.

原子和电子组成了半导体和晶体管，这些正是计算机和互联网的基石。所有这些都可以被 3D 打印。在太空中，塑料垃圾可以变成 3D 打印的材料。

Atoms and electrons make up semi-conductors and transistors, the building blocks of computers and the internet. All can be 3-D printed. In Space, plastic waste can be turned into a 3-D print.

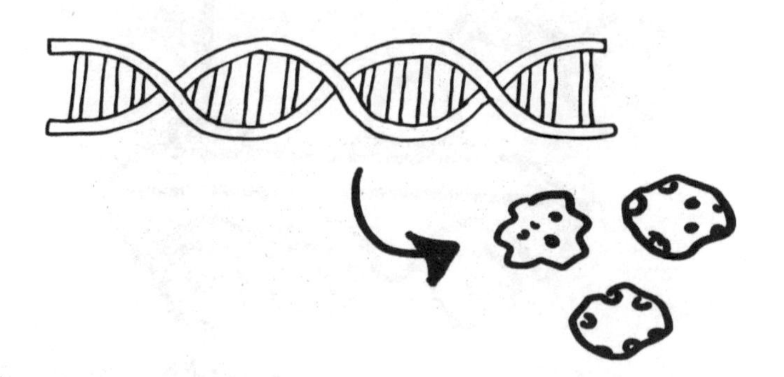

DNA is the one single common molecule of all terrestrial life; its components have been found in meteorites and at the galactic centre of Sagittarius.

DNA 为广大地球生物所共有；它的成分已经在陨石和人马座的星系中心被发现。

Asteroid mining is being planned. As resources on Earth become increasingly scarce, asteroids and comets could become sources of valuable components. These can be sent to Earth or used in Space during exploration.

小行星的开采正在计划中。随着地球上的资源越来越稀缺，小行星和彗星可能会成为有价值成分的来源。这些资源可以被送往地球，也可以在太空探索中使用。

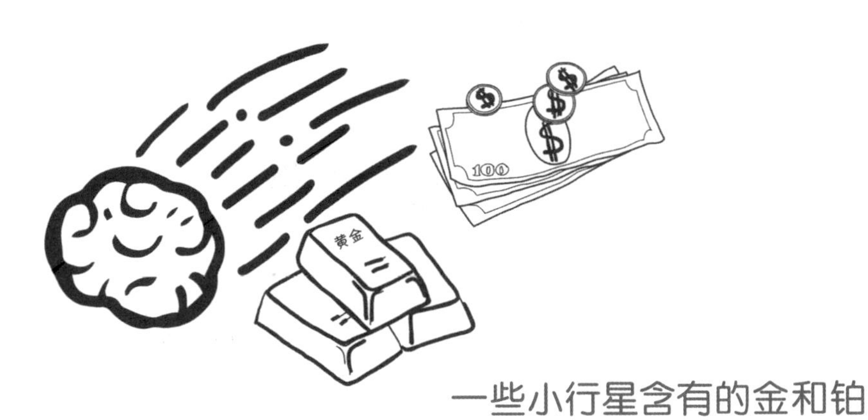

Some asteroids contain more gold and platinum than have been used by people on Earth in their entire history. One asteroid, with the codename UW158, when speeding past Earth was estimated to carry 90 million tons of platinum in its core, valued at US$ 5.4 trillion.

一些小行星含有的金和铂比迄今为止人类在地球上使用过的还要多。在一颗代号为 UW158 的小行星高速掠过地球时，人们估算其核心携带了9 000万吨铂，价值 5.4 万亿美元。

The first welding experiments in Space (with aluminium, titanium and stainless steel) were performed by Russian astronauts, in 1969. The 1973 Skylab mission operated an electric furnace, crystal growth chambers and an electron beam gun, setting up the stage for manufacturing in Space.

1969 年，俄罗斯宇航员进行了第一次太空焊接实验（用铝、钛和不锈钢）。1973 年的太空实验室开动了一台电炉、几个晶体生长设备和一台电子枪，这些工作为太空生产积累了经验。

Would you relocate to go live on Mars?

你会搬去火星上住吗?

Do you think that all we could do has been done on Earth?

你认为在地球上我们能做的一切都已经做了吗?

Could mining asteroids and manufacturing in Space cut down on pollution on Earth?

小行星采矿和太空生产能减少对地球的污染吗?

If you lived on Mars, what would you do differently from what you are doing on Earth?

如果你住在火星上，你所做的事和你在地球上所做的事会有什么不同?

How many people believe that there is life on Mars? Do a survey by asking your friends if they believe that Martians are a reality. Do people who think that there isn't any life on Mars, believe there may be life somewhere else in the Universe? Ask your friends if they would like to travel in Space to experience it for themselves. And is anyone prepared to move to Mars permanently to establish a new society there?

有多少人相信火星上有生命？做一项调查，询问你的朋友是否相信火星人真实存在。那些认为火星上没有生命的人，相信宇宙的其他地方可能有生命吗？问问你的朋友是否愿意去太空旅行体验一下。有人准备永久移居火星，去那里建立一个新社会吗？

学科知识

Academic Knowledge

生物学	在我们的身体和我们的星球，水扮演着生命中至关重要的角色；南极洲湖泊和冰原之下的细菌；木星的卫星；火星表面下发现了液态水；熄火彗星和近地小行星主要由冰组成，可作为长期太空探索任务的水源；甘氨酸是蛋白质中分子量最小的氨基酸，它和DNA前体一起在太空中被发现；量子化学和计算机技术帮助我们理解这些分子的形成。
化　学	构成许多分子的氢、氧和碳原子在宇宙中大量存在；行星和卫星是温暖物质的密集集合，在那里发生着复杂的化学反应；我们所知道的唯一一颗由这种复杂的化学反应产生生命的行星就是地球。
物　理	在宇宙大爆炸中产生了氢和氦，物质在引力作用下聚集成团，形成原始恒星；恒星爆炸时释放的能量产生了元素周期表上剩下的元素；40亿年前，地心引力将重元素拉向地核，留下了一层主要由氧和硅构成的地壳，直到一场小行星雨将金、钴和铁等元素注入枯竭的地壳。
工程学	很多生物物质都可以通过3D打印技术生产出来；用于太空增材制造的材料可以取自小行星；太空生产需要小型化技术、机器人技术、纳米技术、材料科学、增材制造和可再生能源；在太空空间可以生产纯度非常高的材料；用于制药的晶体可以在零重力环境下生长。
经济学	小行星采矿的成本超过了地球上的市场价值，但对月球或火星上的生命来说是必不可少的；一颗直径为1千米的小行星可能含有20亿吨铁镍合金，是全世界产量的3倍；太空探险是高风险的，需要长时间准备和大量投资；从小行星上获取资源比从地球发射物资到太空更有效率。
伦理学	外层空间被定义为“人类的领地”，这使月球岩石的开采合法化，但也禁止人们取得月球土地的产权；1979年的《月球条约》将月球及其自然资源确定为人类共同遗产的一部分；尽管有这项条约，一些企业仍在探讨开发月球资源以获取利润。
历　史	小行星采矿的创意首次出现在1898年的《爱迪生征服火星》中；《外层空间条约》作为外层空间宪法于1967年生效。
地　理	利用三角测量对航天器进行地理跟踪；火星和地球的平均距离为2.25亿千米；太阳和火卫一对火星的潮汐力。
数　学	“水手9号”和“海盗号”轨道飞行器发现火星是非球形的行星体，因此火星的引力势应该用球谐函数描述。
生活方式	外太空生活需要循环利用资源，相关技术也会影响地球上的生活方式。
社会学	外太空的生活依赖于紧密的社会凝聚力以及性格和技能的互补性。
心理学	只有那些意志最坚定的人才会永远离开家园，到别处定居。
系统论	在资源匮乏的环境下，生产和消费将在零排放和零浪费的原则下运行，一切都是可以再利用的；把工业活动转移到太空。

情感智慧
Emotional Intelligence

爷　爷

爷爷向孙女提问，当孙女反问同样的问题时，他的回答非常清晰。他通过承认家族勇敢面对新环境和探索未知的经历来表示他的同理心。他不介意表露自己的情绪。他认为孙女的梦想不可能实现。当孙女坚持她的观点时，他重新表达了自己的看法，坚定地指出她需要科学。他赞扬人类曾经完成的超出预期的壮举。他总结说，这不仅与科学有关，也与态度有关。他阐明火星上的游戏规则非常不同：因为那里没有贫穷，它提供了一个改变游戏规则的机会。然后，他称赞孙女的创造性思维和毅力。

孙　女

孙女好奇心很强，梦想着探索未知的世界。她批评社会缺乏创想。她想在祖先取得的成就上再接再厉，并决心延续家族传统。一想到有机会在火星上开创一片新天地，她就激动不已。她解释了不可能和不太可能之间的区别。她认为，一旦科学发展到了可以发现新事物的阶段，不可能就会变成可能。她分享了自己解决问题的热情，以及在新环境下工作的热情。她很自信，反驳了爷爷提出的质疑。

艺术
The Arts

你想象中的火星人是什么样的？不要被前人的说法所束缚，你可以自由地创造你自己的外星人。想想适应不同的大气成分、温度和重力需要些什么。当你对火星生命与地球生命的不同之处有了清晰的认识，就可以对你想象中的火星人进行相应调整。把你的想法分享给你的朋友和家人，看看他们是否会“喜欢”你创作的火星人。

思维拓展

Systems: Making the Connections

我们还没有成功地解决我们这个时代地球上最紧迫的社会和环境问题。生物多样性的大量丧失，贫困无法消除，难以为每个公民提供高质量的饮用水，无法确保所有人的健康，这些都表明，我们尚未改变我们原有的思维方式。可能会有人告诉我们，人类已经在许多方面取得了进展，但统计数字显示，许多难题并没有得到根本解决。因此需要改变所有的思维方式，在火星上创建第一个社区的设想为我们解决诸多问题提供了一个理想的机会。当你在火星创建一个定居点时，你会有很大的自由度，因为在一个没有生命的环境中设想解决方案，可以设计出新的方法来创造一种可持续的生活方式。人类一直都在探索，而人类历史上探索最为广泛的时期就是一大群人共同踏入未知领域的时候。这种移民文化和探索未知的愿望激励了无数代人在远离其文化和历史中心的地方建立新家园。大多数科学家认为，在火星这样的星球上定居是不可能的，然而，不可能和不太可能之间有明显的区别。借助科学创新，可以把看似不可能的事情变成可能。我们需要的是那些超越前人想象和准备好挑战所有固有思维并致力于探索和寻找持久解决办法的开拓者。这种探索提供了一个时间窗口，在这个窗口中，最先进的科学与社会的迫切需求相结合。总有一些人不受现状束缚并且超越常识。这就是太空定居所代表的，这就是为什么人类在努力探索如何在火星生存的同时，也启发我们该如何改善地球上的生活质量。

动手能力

Capacity to Implement

增材制造又名3D打印，是太空生产的标准模式。探索这种新的制造模式现在对我们地球人来说似乎只是一种创新，但一旦我们离开了这赖以生存的丰饶世界，这种生产方式就不可或缺了。列出三件需要在太空制造的物品。提前设计这些产品。如果你有一台3D打印机，你现在就可以开始学习如何使用它了。请只使用可再生的3D打印材料。

故事灵感来自

This Fable Is Inspired by

阿德里亚娜·马雷

Adriana Marais

阿德里亚娜·马雷拥有南非开普敦大学的理论物理学学士学位和南非夸祖鲁－纳塔尔大学的量子密码学硕士学位。她还拥有量子生物学博士学位。她的博士后研究聚焦光合作用的量子效应和生命起源。她目前正在开普敦大学商学院攻读第二个博士学位，研究资源受限环境下的经济学。阿德里亚娜曾担任软件公司SAP（非洲）的创新主管。她现在是太空发展基金会的主任，该基金会旨在发起非洲的第一次登月任务，并激励发展中国家的年轻人。她是“骄傲人类”的创始人，这是一个致力于开拓前沿技术，从而为地球以及地外星球的可持续发展作贡献的研究网络。

图书在版编目(CIP)数据

冈特生态童书.第七辑:全36册:汉英对照/
(比)冈特·鲍利著;(哥伦)凯瑟琳娜·巴赫绘;
何家振等译.—上海:上海远东出版社,2020
ISBN 978-7-5476-1671-0

Ⅰ.①冈… Ⅱ.①冈… ②凯… ③何… Ⅲ.①生态
环境-环境保护-儿童读物—汉英 Ⅳ.①X171.1-49

中国版本图书馆CIP数据核字(2020)第236911号

策　　划 张　蓉
责任编辑 祁东城
封面设计 魏　来李　廉

冈特生态童书
去火星生活?
[比]冈特·鲍利　著
[哥伦]凯瑟琳娜·巴赫　绘
李原原　译